Filtered Power: Best Practices

Alex Ndukwe

First printing, 2020

Printed in the United States of America

ISBN: 9798692535184

Dedication

My Spiritual Father, **Pastor David Adewuyi**, Assistant Regional Pastor/Province Pastor FCT1, I have known him afar, but it is a great pleasure to have him close to us now.

Forward

Renewable energy is a lucrative business all over the world and Nigeria is not an exception. Everyone wants to be involved to make money, quacks are on the rise.

Most countries have frameworks that aids its regulation, this transcends to choose of materials used for deployments. Embracing good practice ensures that deployed infrastructure would stand a test of time and downtime minimized or eliminated.

The book discusses issues that we are conversant with, we must understand reasons why we choice a particular gadget, taking a deep look at its technology.

Managing associated components are very essential, an example are batteries that stores energy. Sulphation clearly explained and how to resolve such issues is discussed in detail, this would help in the good health of batteries and extended lifespan.

There are diverse Tools and Kits for managing Filtered power solution, two tools were discussed, PowerGuide 4400 and Dewesoft high precision power analyser, they are the best in the market. Power analyser are quite beneficial, a proactive approach to issues before it happens.

Alex ndukwe

Table of Content

	Page
Best Practices explained	7 -9
Chapter One	
Earthing of building and devices	10 - 18
Chapter two	
Monitoring of devices	19 - 47
Chapter Three	
Infrastructure Health sustained	48 - 68
Chapter Four	
Single-phase vs Three Phase	69 - 79
Chapter Five	
PWM vs MPPT Charge controllers	80 - 94
Chapter Six	
Benefits of Voltage Stabilizer	95 - 104

Best Practices explained

Most times we want to deploy filtered powered Infrastructure in a bid to proffer solutions at our homes and corporate environments. One challenge that is quite common is the cutting of corners to ensure that cost of projects is reduced significantly. The book intends to explain these pitfalls adequately.

In the corporate environments Fixed assets which also includes devices that we call infrastructures have a 4 years life span assigned in the fixed asset Register, unfortunately, the TCO(Total Cost of Ownership) shoots up due to incessant repairs, this could have been averted if the best practices are adopted and executed religiously.

My journey for well over twenty years managing Filtered power infrastructure, the device with the highest incidence happened to be the central UPS, users converted this device to standby generators, once there is an outage, they leave the load on the batteries, the implication is that the batteries were replaced within one year and this increased the TCO, naturally the batteries were supposed to last for 3 years. In this case, the best practice is to shut down the load and UPS after 5 minutes if the alternate power, Generator, or city power does not come on.

Quality of the raw power is another important issue, most times there is Earth-leakage without the equipment owner been aware until havoc is done, fire fighting becomes the next line of action, this could have been prevented before it happens.

Power Audit is an important aspect that has been neglected because of the cost of such exercise, innovations have made it possible for the deployment of IoT enabled audit that will replace the rigorous Audit exercise that would have been carried out manually and the cost is also reduced or eliminated.

The filtered powered solution has become especially important in our lives, strategies must be adopted to ensure it keeps working efficiently and its intended objectives are attained or sustained. The book intends to discuss these identify best practices.

Chapter One

Earthing of building and devices

This is very important, the building must be properly earthed, the devices must be earthed, this prevents the destruction of the installed infrastructure like UPS, Inverters, Charge controllers, Solar panels etc. when there is an earth-leakage this implies that there is current on the Neutral(**N**), Earth(**E**) and Live(**L**) of electrical installation.

Earth leakage occurs during the raining season before I explain the process of Earthing a building and devices, our experience at our customer's site, a commercial bank branch. The Network modem belonging to a service provider was shut down by the installed surge protector, the branch had to rely on the VSAT connectivity that was not reliable. After four months, my firm **Tekville systems** was invited for resolution of this leakage, a thorough Audit exercise was carried out by my team, it was now narrowed down to the source of power, the location used the Generator set more frequently, it was not earthed, we ensured that it was earthed and the leakage was arrested. The service provider sent an email to me and demanded the readings because they discovered that their device came on. Can you imagine the pain they have gone through by not adopting best practices?

Earthing of a building

The building would be earthed by creating a central point that would be referred to as 'System', a pit dug around the building, it should be 6 feet. The following would be placed in the pit and they are as follows, earth mat, galvanized earth rods, salt, Earth terminal, charcoal, earth copper cable and thunder arrestor. Let us identify the requirements pictorial and the required steps would be explained afterwards.

Earth Mat

galvanized earth rods

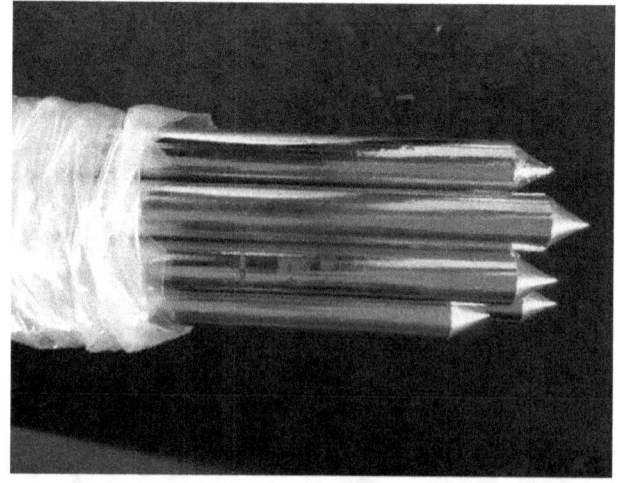

charcoal

Salt

Earth Cable

Earth Bar

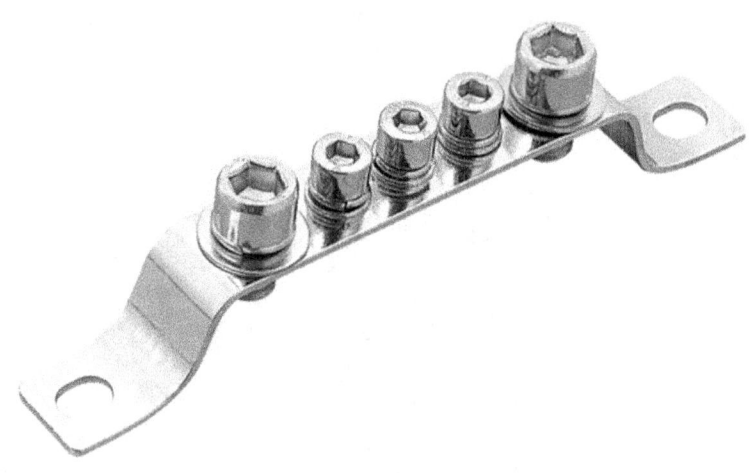

Thunder Arrestor

Connector

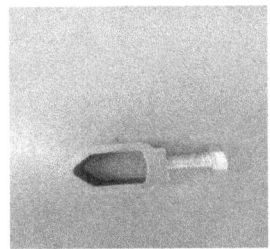

A 6 Feet deep pit needs to be dug, as indicated in the diagram below

The earth mat positioned as shown in the picture, the earth rods placed at the edge of the pit, the mat would be connected to copper cables and Furse connectors to ensure that it is fixed to the ground, let us look at the next diagram.

The earth cable would be connected to the mat and taken outside the pit, charcoal and salt would be sprinkled on the mat as shown in the next diagram

The rods are inserted at the edge of the pit, we will need like 6 units and the pit is covered, the other end of the cable is terminated on the Thunder arrestor hung on the roof. This implies that the building is earthed against leakage, thunderstorms trigger static electricity that could destroy equipment. Each device in the environment are to be earthed, we can create another system, the earth cable would be terminated to the earth bar, each device would be connected to the bar individually, no leakage can affect the equipment.

Chapter two

Monitoring of devices

For efficient performance of filtered power infrastructure, effective monitoring of these installed devices is particularly important. For the case of UPS, inverters, the raw power which is the input current, Output current, Frequency, the temperature can be monitored with a configured SNMP(Simple Network Management Protocol) device attached to equipment and connected to the Network.

This implies that the device would have an IP address, monitoring would be via the web. I could sense that users might not see the need for this and could be stressful, the other question is what if the owner of this equipment is not a technology enthusiast then what happens? The advent of IoT, Internet of Things is adopted to Things like healthcare, retail, manufacturing, energy, logistics. IoT applications in the energy sector grow special attention from consumers, businesses and even governments. Apart from numerous benefits to the electric power supply chain, IoT energy management systems give way to new smarter grids which promise unprecedented savings, improved security, and enhanced efficiency. I am going to recommend IoT enabled devices that can aid the monitoring of Filtered power equipment's and ensure its performance is maximized. Let us look at the SNMP device for UPS monitoring before proceeding with other tools and devices that can be of help.

SNMP ENABLED UPS MONITORING

SNMP allows either one or several UPS systems to be monitored by a computer at a central point on a network. The severity and sensitivity of any alarms can be programmed to alert maintenance personnel direct. This feature is particularly useful when an alarm occurs outside normal working hours. As the software is dealing with actual measurements from a UPS system, and not a Building Management System's interpretation of the operation of a relay contact, the information is far more accurate.

The operating software at the host monitoring computer allows an operational history to be established on each UPS that is monitored. This can be useful in determining loading issues and supply voltage problems without the necessity of connecting expensive monitoring equipment to check an individual UPS system.

The SNMP is purely a monitoring tool to allow the UPS user to monitor the performance of his UPS system, using the host computer's storage capabilities which can hold far more information on the UPS system than the individual UPS. The normal UPS history storage capacity is limited and only allows an engineer to diagnose any potential problems based on recent events. By using the SNMP connection to a host computer, it is possible to see the trend of events over a far greater period.

The monitoring of the UPS systems over the client's network utilizing SNMP can allow the client to decide the severity of any faults that occur to his UPS system and control the expense of calling an engineer to deal with it.

For example, if one module of an N+1 parallel system fails in the middle of the night, the client can make the judgement that he still has protected power and can call an engineer during normal office hours at far less cost than if he responded to the report the UPS system was in alarm from the interpretation of an alarm contact. Correctly used the monitoring of UPS systems by SNMP can be an immensely powerful diagnostic tool and give the user greater knowledge to forewarn of possible issues before they arise.

The manufacturer of the equipment includes the SNMP card on the UPS or a module with a default IP address and the configuration connects the equipment to the network. It is recommended for the environment with Technology personnel to carry out this function of monitoring the device via SNMP.

The Tools in the market are enormous, I will only discuss two products that are considered high tech compared to others, they are as follows:

1) PowerGuide 4400 Power Quality Analysers
2) High-precision Power Quality Analyzer and Meter

PowerGuide 4400 Power Quality Analysers

PowerGuide 4400 Power Quality Analysers are equipped with 8 independent channels and are the only 3-phase advanced power monitor to have a built-in colour touch screen. Automated setups provide instant detection of circuits and configurations, ensuring that the instrument is ready to successfully collect data.

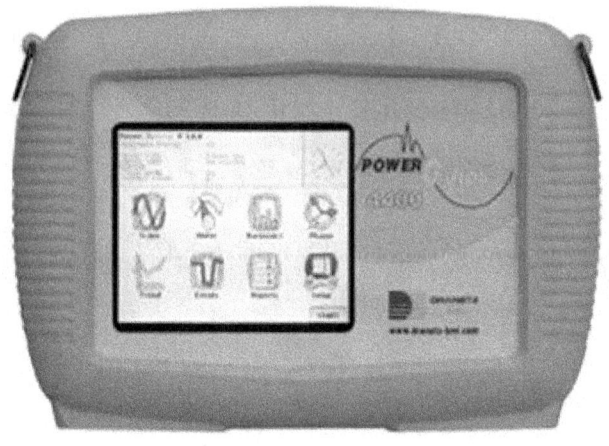

Users can select the length and mode of data collection, including troubleshooting, data logging, power quality surveys, energy, and load balancing. The PowerVisa mains monitor collects data at 256 samples/cycle/channel, offers RS-232, Ethernet or USB remote communications. Data can be viewed in real-time using the scope mode, meter mode, event mode, harmonics spectrum of phasor diagrams. With the touch of a finger or a stylus, data can be viewed and zoom-in on captured disturbances for more detail and automated event characterisation.

Connectivity Solutions

For an effective power continuity plan, it is paramount to install monitoring and control. This can range from UPS monitoring software packages and the appropriate cables and interface cards to a complete Data Centre Infrastructure (DCIM) package.

At EcoPowerSupplies they provide a complete range of software and interface solutions and can assist you in the design and deployment of your chosen monitoring solution. The software packages are available as website downloads and CD-ROM packages and we also supply the appropriate interface adapters (RS232, USB, and volt-free contact cables). Also provides a complete 24/7 remote monitoring service option for its Platinum level maintenance plans and to be sold as a separate service for sites that require high-level of remote monitoring.

Environment Monitoring Systems

Environment Monitoring systems from EcoPowerSupplies include sensors for data centres, comms rooms, server rooms and critical IT facilities where it is important to monitor environmental factors. Environment monitoring systems range from single room monitors for temperature and humidity to site-wide systems for deployment within data centres with multiple server racks. Monitoring solutions allow many parameters to be monitored including power (current, voltage, kW, power factor), temperature ambients, humidity, access, security, water, and smoke. Data centre monitoring systems provide the data required for accurate calculation of your energy usage and overall efficiency calculations including Power Usage Effectiveness (PUE).

The iMeter can be installed as a retrofit to existing basic power distribution units without any smart metering.

High-precision Power Quality Analyzer and Meter

The Dewesoft power quality analyser can measure all the power quality parameters according to IEC 61000-4-30 Class A. In comparison to other power quality meters, Dewesoft can do more detailed power quality analysis. Raw data storing, behaviour at faults, calculation of additional parameters, etc.

Main Features

- **HARMONICS UP TO 150 kHz:** Measure and analyze harmonics for voltage, current, and total harmonic distortion (THD) with frequencies up to 150 kHz. All measurements are carried out according to the IEC-61000-4-7 standards.

- **THD CALCULATION:** The calculation of THD (overall harmonic content) for voltage and current up to the 3000th order.

- **INTERHARMONICS & HIGHER FREQUENCIES:** Measure and analyse interharmonics and higher frequencies. The higher frequency elements can be grouped in 200 Hz bands up to 150 kHz.

- **FLICKER, FLICKER EMISSIONS & RVCs:** Automatic flicker and flicker emission parameters calculation according to the IEC-61400-4-15 and IEC-61400-21 standards.

- **GREAT REAL-TIME VISUALS:** Fast and customizable visual displays are available for FFT, Harmonic FFT, Waterfall FFT with real-time visual feedback making this solution a great power quality monitor.
- **HIGH-SPEED RAW DATA STORING:** Hardware and software provide a powerful storing engine with a continuous stream rate of more than 500 MB per second. Raw data is always stored no matter what. This is to ensure that offline post-processing functions seamlessly whenever it is needed.

- **ADVANCED ONLINE AND OFFLINE MATH PROCESSING:** Dewesoft X data acquisition software offers an easy-to-use mathematics engine. Math can be applied in real-time during measurement or added in post-processing.

- **FAST REVIEW OF DATA:** Datafiles, even if they are gigabytes large, can be opened and reviewed in mere seconds.

- **HIGH-CHANNEL COUNT SOLUTIONS:** We offer large channel count DAQ system configurations with thousands of channels using network configuration as well as more compact portable power quality analyzers for field use.

- **NO HIDDEN COST SOFTWARE LICENSING:** Our software licensing is very flexible and has no renewal or upgrade fees. Upgrades for Dewesoft X DAQ software are FREE forever. You also don't require any additional license to view/analyze the data. Once the data file is stored it can be reviewed and analyzed on unlimited computers without the need for additional software licenses.

Quality Overview

The different power quality parameters describe the deviation of the voltage from its ideal sinusoidal waveform at a certain frequency. These deviations can lead to disturbances, outages, lower power factor, or damages of electrical equipment connected to the grid.

It is essential to permanently track these parameters: starting during the development phase (of the electrical equipment), up until the live operation and beyond: e.g. continuous monitoring of a couple of points in the electrical grid to prevent and correct quality disturbances.

The Dewesoft solution can measure all these parameters according to the IEC 61000-4-30 Class A standard. In comparison to conventional power quality analyzers it is possible to do more detailed analyses (e.g. raw data storing, behaviour at faults, calculation of additional parameters, etc.).

Dewesoft Power Quality Analyzer

The Dewesoft Power Quality Analyzer is a very flexible DAQ solution that combines **power and energy loggers** as well as several other measurement instruments into a single device. This has multiple advantages for the measurement process:

- **Data synchronization:** data is fully synchronized and compatible for comparison.

- **Raw data logging:** raw data is always stored and can, therefore, be analyzed at any time in post-processing.

- **Easy to use:** intuitive software that can perform all the measurement and analysis tasks, meaning the learning curve is much easier to master.

- **Low cost:** a single instrument that can measure and analyze power parameters that would usually require multiple instruments will save space, time, and money.

All this combined makes the Dewesoft power quality analyzer an indispensable piece of equipment that acts as an arsenal of devices able to measure among others:

- Harmonics and THD up to 150 kHz

- Interharmonics & higher frequencies

- Flicker, Flicker Emissions, RVCs
- FFT, Harmonic FFT, Waterfall FFT
- Symmetrical components

Power Quality Standards Overview

power quality meters fulfil all the requirements of power quality standards and can be used in a wide range of testing applications. The table below sums up the supported power quality standards:

IEC 61000-4-30	Requirements for Power Quality
IEC 61000-4-7	Analyzers, Calculation
IEC 61000-4-15	of Harmonics, Flicker, etc.

Standard	Description
EN 50160 EN 50163 IEEE-519 IEC 61000-2-4 etc.	Power Quality limits of the public grid, industries, and railway applications.
IEC 61400-21 IEC 61400-12 FGW-TR3 BDEW VDE-AR4105 etc.	Power quality analysis of renewable energy sources.
IEC 61000-3-3 IEC 61000-3-11	Electromagnetic compatibility (EMC) of voltage changes and Flicker.
IEC 61000-3-2 IEC 61000-3-12	Electromagnetic compatibility (EMC) of harmonics current.

Harmonics calculations

- U, I, P, Q, and impedance
- Individual setup of the number of harmonics including DC component (Example: 20 kHz sampling rate = 200 harmonics @ 50 Hz)
- Harmonics up to 3000th order (@50 Hz)
- Variable sidebands and half sidebands for Harmonics
- Higher Frequencies up to 150 kHz in 200 Hz bands
- Interharmonics, groups or single values
- According to EN 61000-4-7
- Calculation corrected to the actual real frequency
- THD, THD even, THD odd
- Trigger on each parameter
- Background harmonics subtraction

FFT Harmonics Analysis

Harmonics are integer multiples of the fundamental frequency (e.g. 50 Hz) and distort voltage and current of the original waveform. Harmonic voltages and currents caused by non-sinusoidal loads can affect the operation and lifetime of electrical equipment and devices.

Harmonic frequencies in motors and generators can lead to increased heating (iron & copper losses), can affect torque (pulsating or reduced torque), create mechanical oscillations, and higher audible noise, which can cause ageing of shaft, insulation and mechanical parts and reduce the efficiency.

Current harmonics in transformers increase copper and stray flux losses. Voltage harmonics increase iron losses. The losses are directly proportional to the frequency and, therefore, higher frequency harmonic components are more important than lower frequency components. Harmonics can also cause vibrations and higher noise. The effects of other electrical equipment and devices are similar and are mainly reduced efficiency and lifetime, increased heating, malfunction or even unpredictable behaviour.

Harmonics, Interharmonics and THD

The Dewesoft solution can measure harmonics for voltage, current and additional active and reactive power up to the 3000th order. All calculations are implemented according to **IEC 61000-4-7 standards**.

You can define the number of sidebands and half-bands for the harmonic order calculation. The higher frequency parts can be grouped in 200 Hz bands up to 150 kHz.

The calculation of THD (overall harmonic content) for voltage and current up to **3000th order** and the interharmonics complete the analysis functionalities.

These powerful harmonic calculation functions offer analysis for all types of electrical equipment and devices.

Harmonics calculations

- U, I, P, Q, and impedance
- Individual setup of the number of harmonics including DC component (Example: 20 kHz sampling rate = 200 harmonics @ 50 Hz)
- Harmonics up to 3000th order (@50 Hz)
- Variable sidebands and half sidebands for Harmonics
- Higher Frequencies up to 150 kHz in 200 Hz bands
- Interharmonics, groups or single values
- According to EN 61000-4-7
- Calculation corrected to the actual real frequency
- THD, THD even, THD odd
- Trigger on each parameter
- Background harmonics subtraction

FFT Harmonics Analysis

Harmonics are integer multiples of the fundamental frequency (e.g. 50 Hz) and distort voltage and current of the original waveform.

Harmonic voltages and currents caused by non-sinusoidal loads can affect the operation and lifetime of electrical equipment and devices.

Harmonic frequencies in motors and generators can lead to increased heating (iron & copper losses), can affect torque (pulsating or reduced torque), create mechanical oscillations, and higher audible noise, which can cause ageing of shaft, insulation and mechanical parts and reduce the efficiency.

Current harmonics in transformers increase copper and stray flux losses. Voltage harmonics increase iron losses. The losses are directly proportional to the frequency and, therefore, higher frequency harmonic components are more important than lower frequency components. Harmonics can also cause vibrations and higher noise. The effects of other electrical equipment and devices are very similar and are mainly reduced efficiency and lifetime, increased heating, malfunction or even unpredictable behaviour.

Harmonics, Interharmonics and THD

The Dewesoft solution can measure harmonics for voltage, current and additional active and reactive power up to the 3000th order. All calculations are implemented according to **IEC 61000-4-7 standards**.

You can define the number of sidebands and half-bands for the harmonic order calculation. The higher frequency parts can be grouped in 200 Hz bands up to 150 kHz.

The calculation of THD (overall harmonic content) for voltage and current up to **3000th order** and the interharmonics complete the analysis functionalities.

These powerful harmonic calculation functions offer analysis for all types of electrical equipment and devices.

Harmonics calculations

- U, I, P, Q, and impedance
- Individual setup of the number of harmonics including DC component (Example: 20 kHz sampling rate = 200 harmonics @ 50 Hz)
- Harmonics up to 3000th order (@50 Hz)
- Variable sidebands and half sidebands for Harmonics
- Higher Frequencies up to 150 kHz in 200 Hz bands
- Interharmonics, groups or single values
- According to EN 61000-4-7
- Calculation corrected to the actual real frequency
- THD, THD even, THD odd
- Trigger on each parameter
- Background harmonics subtraction
-

Full FFT Frequency Spectrum Analysis

In addition to the FFT harmonics, a full frequency-based FFT analysis is available. All frequencies can be analyzed with this function Trigger on FFT patterns and offer definable filters:

- hanning,
- haming,
- flat top,
- rectangle,
- etc.

FFT Waterfall Analysis

Besides the FFT and the harmonic FFT analysis, the Power Quality Analyzer also offers a **2D and 3D FFT waterfall analysis** option.

This type of data visualization is especially useful for the analysis of variable drives. For example, looking at a run-up of an inverter, it is visible how the harmonic sidebands originate as the frequency increases. The image depicts the runup of an inverter of a traction drive from 0 to 150 Hz.

The FFT waterfall visual display can be linear or logarithmic, 2D, or 3D and sorted by harmonic order or frequency.

Flicker Test and Flicker Emission Test

Flicker is a term used to describe fluctuations (repetitive variations) of the RMS voltage between two steady-state conditions. Flashing light bulbs are indicators for high flicker exposure. Flicker is especially present in grids with a low short-circuit resistance and is caused by the frequent connection and disconnection (e.g. heat pumps, rolling mills, etc.) of loads that affect the voltage. A high level of flicker is perceived as psychologically irritating and can be harmful to humans. The Dewesoft Power Quality analyzers offer the following features for flicker measurement:

- Measure all flicker parameters according to the **IEC 61000-4-15 standard**.
- Flicker emission calculation according to the **IEC 61400-21 standard** and allows the evaluation of flicker emission into the grid caused by wind power plants or other generation units.
- PST and PLT with flexible intervals.
- Individual recalculation intervals.
- P_{inst}, d_U, d_{Umax}, $d_{Uduration}$

Rapid Voltage Changes

Rapid Voltage Changes (RVCs) are parameters which are added as a supplement to the flicker standard. Dewesoft X data acquisition software calculates these parameters according to the **IEC 61000-4-15 standard**.

Rapid Voltage Changes describe any voltage fluctuations which change the voltage amplitude between two steady states more than 3% for a certain time interval. These voltage changes can be analyzed in post-processing using various parameters:

- the depth of voltage change,
- d_u, d_{max}, $d_{uduration}$,
- steady-state deviation,
- Etc.

Unbalance - Symmetrical Components

A balanced system has a **120° phase shift** between the voltages and currents, and the voltages and currents have the same amplitude respectively. **Unbalance** happens when the 3 phase system is loaded unevenly, and the phases and magnitudes no longer correlate.

To analyze an unbalanced system, the **symmetrical components calculation method** is used. This method splits the original unbalanced 3 phase power system into a positive system (rotation like the original system), the negative system (rotation in the reverse direction), and a zero system.

An unbalanced system could cause current to flow in the neutral line, overheating electrical components, mechanical stress, increased vibration, and torque pulsation, low power quality, and energy losses among others.

Dewesoft's Power Meters can measure over 50 parameters for a comprehensive analysis of an unbalanced system. These parameters include various calculations for voltage, current, active-, reactive-, apparent power as well as harmonics.

Frequency Deviations

The Dewesoft Power Analyzer can be used for frequency monitoring and testing frequency behaviour of power generation units in the developmental stage (please see renewable energy testing).

High-frequency deviations from the **fundamental frequency** in public grids can have severe consequences. If the frequency drops or rises excessively there is a chance that the whole power system could collapse causing a blackout.

Frequency deviations in power grids are caused by the connection and disconnection of power generation plants or big loads. The grid becomes unstable if there is a deviation from the nominal frequency at which the grid operates. If the frequency is too high, there is too much power in the grid due to power oversupply. If the frequency is too low, there is too little power in the grid due to power undersupply. As renewable energies especially, wind and solar become more popular, grid stability is more at risk than ever. This is since wind does not always blow at constant speeds and that sun energy gets impacted by clouds, shadows, and fluctuations in radiation intensity. This leads to abrupt deviations in the frequency at which power is delivered to the grid.

Chapter Three

Infrastructure Health sustained

The preceding chapter we showcased two important tools that could help in monitoring installed filtered power infrastructure before the issue occurs information or data is made available for appropriate decision to be taken.

There are some other issues we must take seriously in the quest for the sustained health of installed power infrastructure, in Nigeria this sector is not regulated and this attribute for reasons we have lots of substandard products and accessories in the market. In my book on users guide I mentioned a couple of issues that need to be addressed when deploying solar power.

For sustained good health, we must pay attention to the following:

1) Quality accessories Must be used during installation
2) Deal with sulphation from Inception
3) Do not install infrastructure if the building is not earthed
4) Make sure the equipment are earthed
5) Ensure the Design is correct from day 1
6) Do not patronise quacks
7) Avoid Overloading
8) Ensure Alarms on equipment is resolved
9) Avoid cutting corners
10) Understand the products your purchase

Quality accessories Must be used during installation

Installing Filtered power infrastructure depends on the desired solution, even at that we need to use accessories like **cables** of various sizes, **circuit breakers**, **Charge controllers**, **batteries** etc. From experience, Nigerian cables are the best in terms of quality and the prices are higher than Chinese made, poor quality cables would not allow the deployed solution to work properly.

The Standards Organisation of Nigeria (SON) has raised the alarm over what it described as the presence of substandard SUNRISE brand of electric cables in the Nigerian markets and warned consumers to avoid patronising the product. The destruction, according to the statement, was consequent upon due Conformity Assessment, the results of which showed that the SUNRISE Cable has a very high conductor resistance, poor conductor elongation and very low tensile strength of insulation against the requirements of NIS IEC 60228 for a conductor, NIS IEC 60227 series and NIS IEC 60811. It thus poses a grave danger of heating up which could easily lead to fire outbreaks, according to SON.

(SON raises the alarm over circulation of substandard electric cables in Nigeria, SON Newsletter, July 2020)

Aboloma showed to everyone present that the importer cloned over 21 branded cables include Nexams Nigerian wire, Kable mek, Pure chem, Coleman cables, Nocan, Kb Cables, Niger Chem, and other unbranded ones, among others. He added that the agency had made some progress in the area of sanitising the nations cable market, adding that dubious importers were trying to cash in on that achievement by going overseas to clone Nigerian cables adjudged to be best in the world.

(Ending Influx of Substandard Cables in Nigeria, Thisday Newspaper, June 27, 2017)

We must ensure quality cables are used when deploying a filtered power infrastructure, apart from guarantying quality installation, the health of the equipment or device is sustained.

Circuit breakers are designed for protection, it trips off when electrical issues like high voltage occur. Installing quality breakers would ensure that its functions are performed when required. Breakers are attached within a battery bank; these are for protection purposes. The market is flooded with all kinds of devices, quality must not be compromised.

We will also notice that the DB, distribution box is equipped with circuit breaker this protects devices in the building if it is not effective valuable appliances would be lost, let us look at common electrical issues in a building:

FREQUENT ELECTRICAL SURGES

Electrical surges can be caused by anything from lightning strikes, damage to power lines, faulty appliances, and bad electrical wiring in the house. While an actual surge only lasts a microsecond, frequent surges can damage the electrical components connected to your home, degrading their life expectancy significantly. If you experience frequent electrical surges, the culprit is probably an electrical device connected to the home grid or the wiring itself. Try removing any cheaply made devices or power boards from the outlet to see if this prevents the surges. Otherwise, it might be time to consult a professional electrician.

SAGS AND DIPS IN POWER

Like electrical surges, sags and dips in electrical supply can often be attributed to devices connected to your power grid that are faulty or made with substandard materials and draw a lot of power when they are turned on.

LIGHT SWITCHES NOT WORKING PROPERLY

Dimmer switches that do not adjust light properly can often be attributed to shoddy workmanship or sub-standard products. If you have just moved into a new house and find switches that don't seem to activate anything at all, this might be a sign the switches have been superseded and fixtures removed, or it could be a fault in the outlet, circuit or wiring. Consult with an electrician if you are experiencing issues with switches in your house. Want some electrical safety tips for your home?

CIRCUIT BREAKER TRIPPING FREQUENTLY

High wattage items like microwaves and hairdryers can trip circuit breakers, particularly when other power-consuming items are used on the same source. A circuit breaker is designed to protect you and your home, so when it does trip, that's a sign it's doing its job.

Look at what you were using when it tripped. If it was a hairdryer, try using the low setting. Alternatively, limit the electrical usage on a single circuit while high watt devices are in use.

CIRCUIT OVERLOAD

One of the biggest causes of frequent circuit breaker tripping is the overloading of power boards. Most homes and apartments, even newer ones, don't have enough power points to cater to, for example, a complete home entertainment unit setup. If circuit breakers in your home are tripping frequently, it could be down to circuit overload. Prevent this by:

Never daisy-chain power boards.

Remove devices that aren't in use (for example, phone chargers still draw power even when not connected).

Spread your electrical needs around. Don't overburden a single circuit.

Be mindful of how you connect devices around the home – what's in use, and what is unnecessary.

LIGHTS TOO BRIGHT OR DIM

If some lights around the house seem excessively bright but others are dim, then there are two probable causes:

Different types of lights with different wattage: Check that all the globes are identical.

Bad main neutral connection: This will continue to cause problems for the home until it is fixed by a professional.

ELECTRICAL SHOCKS

An electrical shock is a nasty experience. Even though they are usually pretty mild, something akin to a static shock, they remind us that electricity is dangerous when not probably utilised.

Electrical shocks typically happen when you turn a device on or off. The issue could be with the appliance, or it could be in the wiring. You can test this by plugging in another device and seeing if the results are reproducible, however, you're just risking another electrical shock. In most cases, it might be better to speak with an electrician.

LIGHT BULBS BURNING OUT TOO OFTEN

There are several reasons your lights can be burning out too often:

- Wattage is too high

- Insulation is too close to the light
- Bad wiring on the circuit
- Bad wiring on the mains

On a dimmer switch, too much total wattage on one switch

If flickering, there is probably a poor connection on the circuit.

Isolating the issue can be tricky for non-professionals. If you are going through light bulbs like it's nobody's business, it might be worth reaching out to an electrician to help identify the root cause of light bulb burnouts.

RECESSED LIGHT 'GOES OUT' AND COMES BACK ON

Recessed lighting (like downlights) are equipped with safety devices that cut out power to the light when it gets too hot. You're either using too high wattage on the bulb, or insulation in the ceiling is too close to the bulb.

When such issues are experienced, installed UPS might not charge correctly and the batteries are at the risk of sulphation due to inadequate charge. Except the deployment is an off-grid solution.

Deal with sulphation from Inception

The battery is the most neglected component of filtered power; its maintenance is usually neglected. What is battery sulphation? During use, small sulphate crystals form, but these are normal and are not harmful. During prolonged charge deprivation, however, the amorphous lead sulphate converts to a stable crystalline and deposits on the negative plates. This leads to the development of large crystals that reduce the battery's active material, which is responsible for the performance.

There are two types of sulfation: reversible (or soft sulfation), and permanent (or hard sulfation). If a battery is serviced early, reversible sulfation can often be corrected by applying an overcharge to an already fully charged battery in the form of a regulated current of about 200mA. The battery terminal voltage can rise to between 2.50 and 2.66V/cell (15 and 16V on a 12V monoblock) for about 24 hours. Increasing the battery temperature to 50–60°C (122–140°F) during the corrective service further helps in dissolving the crystals.

Permanent sulfation sets in when the battery has been in a low state-of-charge for weeks or months. At this stage, no form of restoration seems possible; however, the recovery yield is not fully understood. To everyone's amazement, new lead-acid batteries can often be fully restored after dwelling in a low-voltage condition for many weeks. Other factors may play a role. A subtle indication of whether lead-acid can be recovered or not is visible on the voltage discharge curve. If a fully charged battery retains a stable voltage profile on discharge, chances of reactivation are better than if the voltage drops rapidly with a load.

Several companies offer anti-sulfation devices that apply pulses to the battery terminals to prevent and reverse sulfation. Such technologies will lower the sulfation on a healthy battery, but they cannot effectively reverse the condition once present. It's a "one size fits all" approach and the method is unscientific. Applying random pulses or blindly inducing an overcharge can harm the battery by promoting grid corrosion. There are no simple methods to measure sulfation, nor are commercial chargers available that apply a calculated overcharge to dissolve the crystals. As with medicine, the most effective remedy is to apply a corrective service for the time needed and not longer.

While anti-sulfation devices can reverse the condition, some battery manufacturers do not recommend the treatment as it tends to create soft shorts that may increase self-discharge.

Furthermore, the pulses contain ripple voltage that causes some heating of the battery. Battery manufacturers specify the allowable ripple when charging lead-acid batteries.

prevenling permanent sulfation?

The best way to prevent sulfation is to keep a lead-acid battery fully charged because lead sulphate does not form. This can be accomplished in three ways. The best solution is to use a charger that can deliver a continuous "float" charge at the battery manufacturer's recommended float or maintenance voltage for a fully charged battery. 12-volt batteries, depending on the battery type, usually have fixed float voltages between 13.2 VDC and 13.8 VDC, measured at 80° F (26.7° C) with an accurate (.5% or better) digital

voltmeter. Based on the battery type you are using, charging can best be accomplished with a microprocessor-controlled, three-stage (for AGM or Gel Cell batteries) or four-stage (for wet batteries) "smart" charger or by voltage-regulated float charger to "float" or maintain fully a charged battery. A cheap, unregulated "trickle" charger or manual two-stage charger can overcharge a battery and destroy it.

A second and less desirable method is to periodically recharge the battery when the State-of-Charge drops to 80% or below. Maintaining a high State-of-Charge (SoC) tends to prevent irreversible sulfation. The recharge frequency is dependent on the parasitic load, temperature, the battery's condition, and plate formulation (battery type). Temperature matters! Lower temperatures slow down electrochemical reactions and higher temperatures speed them up. A battery stored at 95° F (35° C) will self-discharge twice as
fast than one stored at 75° F (23.9° C).

A third technique is to use a regulated solar panel or wind or water generator designed to float charge the battery. This is a popular solution when AC power is unavailable for charging.

Light Sulfation

Check the electrolyte levels and apply a constant current at 2% of the battery's RC or 1% of the AH capacity rating for 48 to 120 hours at 14.4 VDC or more, depending on the electrolyte temperature and capacity of the battery. Cycle (discharge to 50% and recharge) the battery a couple of times and test its capacity. You might have to increase the voltage to break down the hard lead sulfate crystals. If the battery gets above 125° F (51.7° C) then stop charging and allow the battery to cool down before continuing.

Heavy Sulfation

Replace the old electrolyte with distilled, deionized or demineralized water, let stand for one hour, apply a constant current at four amps at 13.8 VDC until there is no additional rise in specific gravity, remove the electrolyte, wash the sediment out, replace with fresh electrolyte (battery acid), and recharge. If the specific gravity exceeds 1.300, then remove the new electrolyte, wash the sediment out, and start over from the beginning with distilled water. You might have to increase the voltage to break down the hard lead sulfate

crystals. If the battery gets above 125° F (51.7° C) then stop charging and allow the battery to cool down before continuing. Cycle (discharge to 50% and recharge) the battery a couple of times and test capacity. The sulfate crystals are more soluble in water than in electrolyte. As these crystals are dissolved, the sulfate is converted back into sulfuric acid and the specific gravity rises. This procedure will only work with some batteries.

Desulfators

Use a desulfator also known as a pulse charger. A list of some of the desulfator or pulse charger manufacturers are as follows:

- Pulse tech, United States
- Nanjing Longline Electronic Technology Co., Ltd Jiangsu, China
- Gudy Electronics Limited Guangdong, China
- Shenzhen Everexceed Industrial Co., Ltd. Guangdong, China
- HANGZHOU DEKANG INTELLIGENT EQUIPMENT CO., LTD. Zhejiang, China

Despite manufacturer's claims, some battery experts feel that desulfators and pulse chargers do not work any better at removing permanent sulfation than do constant voltage chargers.

Do not install infrastructure if the building is not earthed

The first chapter stressed the importance of earthen, it would be unwise to install infrastructure in a building not earthed, the advice is simple, while the deployment is in progress ensure you arrange for the earthen to be carried out, it won't cost much. The thunder arrestor might be the convectional one made of copper to reduce cost. The potential danger is that the equipment might be damaged or lost.

Make sure the equipment are earthed

It is a good practice to ensure that the devices like inverter, stabilizers, any other associated device are earthed. This is a safety measure to ensure the devices are protected in case there are issues with the earthing of the building or leakage occurs accidentally.

Ensure the Design is correct from inception

Design of Pv system must be correctly done from inception. The Equipment purchased must meet requirements with reference to the Load to be powered, filtered power points must be separated from the Mains, clearly labelled to avoid overloading. The choice of charge controller correctly specified. Our intended goal determines what the design would look like, professionalism must be maintained and not compromised.

Do not patronise quacks

When an issue arises it is advisable to consult professionals and not quacks, you could lose your equipment in the process, increase the days of downtime or be pushed to replace the equipment without notice. Electricians don't understand the repairs of inverters, Stabilizers etc. Consult competent vendors in the line of business.

Chapter Four

Single-phase vs Three-Phase Power

Phase refers to the distribution of a load. What is the difference between single-phase and three-phase power supplies? Single-phase power is a two-wire alternating current (ac) power circuit. Typically, there is one power wire—the phase wire—and one neutral wire, with current flowing between the power wire (through the load) and the neutral wire. Three-phase power is a three-wire ac power circuit with each phase ac signal 120 electrical degrees apart.

Residential homes are usually served by a single-phase power supply, while commercial and industrial facilities usually use a three-phase supply. One key difference between single-phase vs. three-phase is that a three-phase power supply better accommodates higher loads. Single-phase power supplies are most used when typical loads are lighting or heating, rather than large electric motors. Single-phase systems can be derived from three-phase systems.

Do you know that a single-phase power brings electricity to your residential home? Why do you think most people prefer three-phase power than a single-phase power?

A phase is the current or the voltage among an existing wire as well as a neutral cable. Its waveform of electronic signals will be analyzed by an oscilloscope, a digital device that draws a graph showing the instantaneous signal voltage as a function of time.

Further, a phase depends on its electric load distribution corresponding to a type of unit, whether a single-phase or three-phase power.

A single-phase (1-phase) has less power, requiring two wires; while, three-phase (3-phase) requires more, including three or four wires.

What is Single-Phase Power?

Single-phase power simultaneously changes the supply voltage of an AC power by a system. More often, single-phase power is known as "residential voltage," since it is that most homes use. In the distribution of power, a single-phase uses the phase and neutral wires. Phase wire carries the current load, while the neutral wire provides a path where the current returns. It creates a single sine wave (low voltage). The common voltage for a single-phase power starts at 230V. Also, its frequency approximates to 50Hz.

Single-phase motors require extra circuits to work since a single-phase supply connecting to an AC motor doesn't generate a rotating magnetic field. The power output of a single-phase supply is not constant, meaning its voltage supply rises and falls.

What are the advantages of using Single-Phase Power?

Single-Phase Power generates electricity to residential homes and domestic supplies, since most appliances require only a small amount of power to function, including fans, heaters, television, refrigerator, and lights.

The design and operation are plain and ordinary. It has a lightweight and compact unit, which the current through the line will be less when the transmission of voltage is high.

Due to the reduction of I2R, the current is low. Meaning, single-phase power ensures the unit to operate at optimum with increased efficiency of its transmission. Single-phase power is best to use with fractional, or lower horsepower units up to 5 HP. What are the disadvantages of using Single-Phase Power?

Small single-phase motors need an additional circuitry such as Motor Starters (similar to starter capacitors in fans and pumps) since its single-phase supply is insufficient for an initial start-up. Industrial motors require heavy electronic loads. Ergo, it cannot run on a single-phase supply.

What is Three-Phase Power?

Three-phase power provides three alternating currents, with three separate electric services. Each leg of alternating current reaches a maximum voltage, only separated by 1/3 of the time in a full cycle. In other words, the power output of a three-phase power remains to be constant, and it never drops into zero. In a three-phase power supply, it requires four wires, namely one neutral wire and three-conductor wires. These three conductor wires are 120-degree distant from each other. Also, each AC Power Signal is 1200 out of phase with each other.

Moreover, there are two types of circuit configurations in a three-phase power supply, such as the Delta and the Star. The Delta Configuration requires no neutral wire and only all high voltage systems use it; while Star Configuration requires a neutral wire and a ground wire.

What are the advantages of a Three-Phase Power?

Run larger loads easily. Commercial and Industrial loads prefer a three-phase power supply since it requires more heavy electronic loads.

Do not require any starters to three-phase motors used in big industries, since it has sufficient phase difference to supply initial torque for the motor to start.

The three-phase power supply requires less conducting materials to transmit and distribute electrical power. Hence, it becomes more economical when it speaks about costs. As the number of phases increases in the system, the DC voltage of a three-phased power becomes smoother and more advantageous.

What are the disadvantages of a Three-Phase Power?

Since the system voltage is quite high, the three-phase power supplies and motors maintain a high cost of insulation. Insulation depends on the voltage of the unit, while its size of the wire depends on the current.

Three-phase power units cannot handle the overload. Meaning, when it results in damage, the cost of repair is higher since changing individual components is expensive.

What are the differences between Single-Phase Power and Three-Phase Power?

Required Wirings on Power Supply

In a single-phase power supply, it only requires two wires, namely Phase and Neutral. On the other hand, three-phase power supply only works through three wires, including three-conductor wires and a neutral wire. Thus, the costs of cabling and total installation are both reduced when you deliver three-phase power directly to your server cabinets.

Voltages

In a single-phase power supply, it only suffices to 230V, whereas a three-phase power supply maximizes up to 415V.

Its place of utility

Residential homes usually utilize lower power supply, requiring less quantity of power to function your mobile devices and home appliances. In contrast, commercial and industrial companies require heavier electronic load. Hence, it utilizes a three-phase power supply to function.

Efficiency

A single-phase power cannot start by themselves, requiring external devices such as Motor Startups. As its opposite, a three-phase power can start by itself without requiring any external devices. Also, it can even reverse the directions of two conductors.

Application

A single-phase power supply generates a lower amount of electricity to support homes and non-industrial businesses, whereas a three-phase power supplies power grids, data centres, aircraft, shipboard, and other electronic loads larger than 1,000 watts.

To sum up, everything, choosing between a single-phase power or a three-phase power is a question of your necessity, economy, and practicality. While you benefit from these two power supplies, always consider your practical need. For a real-life application, Circuit Specialist suggests you choose a single-phase power for domestic and residential use.

A single-phase power has simple and ordinary physical features, requiring a small amount of power to function your mobile devices and home appliances.

But, when you have a three to four air conditioning units running at the same time, double door refrigerators, supersize washing machines, then Circuit Specialist advises you to avail a three-phase power supply to distribute each load properly.

Nevertheless, while both single-phase and three-phase powers have palpable differences, you should always consider the following factors to have a wise investment:

- required wirings on the power supply;

- voltages;
- its place of utility;
- the efficiency of performance; and
- application.

Chapter Five

PWM vs MPPT Charge controllers

This chapter allows us to compare PWM and MPPT charge controllers, most times we are at the mercy of our Technician or vendor, they install whatever they wish. I am going to compare MPPT vs. PWM in terms of:

- Working Principle
- Operation
- Performance
- Pricing and value
- Unique features
- Our Recommendation

What a Solar Charge Controller do?

Charge Controller (aka Solar regulator) is a kind of controller regulate the charge and discharge process in the solar power system. the main role of the charge controller is to control the charge current flowing from the PV panels to the battery, keep the current flowing not too large to prevent the battery pack from overcharge. there are 2 kinds of the solar regulator on the market:

- MPPT Solar Charge Controller
- PWM Solar Charge Controller

Parameter	PWM	MPPT
Charge Method	3 Stage	Multi-Stage
Conversion Rate	75% - 80%	99%
Size	20A – 60A	30A – 100A
Scalability	< 2KW	> 2KW
Cost		

MPPT vs PWM: Principle

MPPT and PWM are both energy control methods used by the charge controller to regulate the current flowing from the solar panel to the battery. PWM has a cheap price and a 75% conversion rate, mppt ask a higher price, but the latest MPPT can get a huge conversion rate improvement which up to 99%.

What is MPPT

MPPT Stands for Maximum Power Point Tracking is a technique for tracking and regulate the output energy from the solar panel to the battery.

The MPPT detects the solar panel output voltage and current in real-time and continuously track the maximum power (P=U*I), regulates the output voltage correspondingly so that the system can always charge the battery with the maximum power.

What is PWM

PWM is short for Pulse Width Modulation, it's a technique to modulating the width of pulse under certain rules, thus change the voltage and frequency of output energy from the solar panel to charge the battery. the PWM charge controller can be considered as an electric switch between the solar panel and battery packs.

MPPT vs PWM: Charge Difference

PWM charges the battery with constant 3 stage Charging (bulk, float and absorb), while the MPPT is Maximum power point tracking and can be thought of as multi Stage charging. The conversion efficiency of MPPT is relatively 30% higher compared to PWM.

PMW 3 Stage Charging:

Bulk Charge: the bulk charge stage is when the PV system proceeds the most charge to the solar battery, When the battery voltage is low, the system will charge the battery with high current and voltage. It should be noted that there is a maintenance point (overcharge protection) and when the voltage at the battery end is higher than this maintenance value during charging, the direct charging should be stopped.

Absorb Charge: After the first stage charging, the battery will normally wait for a period to allow the voltage to fall naturally, and then it enters the balanced charging stage. the stage also is called the constant voltage charging.

Float Charge:

Float charge is the last stage of 3 stage charging, as known as Trickle charging, A trickle is a small charge current to a battery at a low rate and in a constant manner. Most rechargeable batteries lose power after being fully charged due to self-discharge. If the charge continues at the same small current as of the self-discharge rate, the charge power can be maintained.

MPPT Multi-Stage Charging:

MPPT also has a 3 stage charging process, unlike PWM, MPPT can auto-switch the charge method base on the PV condition. here is how:

Bulk Charge: at the bulk charge stage, the mppt controller working at Vmpp mode and automatically adjusts the output voltage to charge the battery, this ensures the system harvest the most power from the PV array. unlike PWM, at the bulk charge stage, the output is at a constant value.

When the sunlight is strong, the output power of photovoltaic cells increases greatly, and the charging current can soon reach the threshold, abort the MPPT charging and switch to the constant-current charging method.

When the sunlight becomes weak and hard to held constant current charge, then switch to the MPPT charging method, and so freely switch until the voltage at the battery side rises to the saturation voltage Ur, the battery enters the constant voltage charging stage. By combining the MPPT charging method with the constant current charging method and automatic switching, it is possible to make full use of solar energy to charge the battery quickly.

Balance Charge / Boost Charge: As the battery is charged to the boost voltage setpoint, the solar controller continuously adjusts the charging current to keep the battery charging process.

Absorb Charge: at the absorb charge stage, as the battery voltage increased, the charging current gradually decreases, when the charging current falls to about 0.01C, the constant voltage charging ended.

Float Charge:

The charge voltage (Uf) in the floating charge mode is a little lower than the constant charge. The main purpose of this charging stage is to compensate for the self-losing energy of the battery. at the process, the battery gets fully charged.

MPPT Charging Advantages:

Quickly scan the whole I-V curve and track the PV cell PowerPoint in seconds.

Innovative maximum power point tracking technology can significantly improve the solar system energy utilization rate, conversion efficiency up to 97%.

Increase charging efficiency by at least 20% compared to PWM

With the same load, using the MPPT charging method can reduce the power of PV modules and reduce module costs.

MPPT vs PWM Solar Charge Controller: Application

PWM Solar Charge Controller is the most common, cost-efficient, and easy deployed charge controller solution for small off-grid power systems. MPPT Regulators have much more electronics components inside and are much more complex than PWM type charge controllers.

PWM solar controller usually applied in a small 12V or 24V PV system. MPPT type regulator also can work with 12v, 24v system but some models also can handle 36v and 48v system when needed. size the right ampere for the controller is easy with this tool.

MPPT Charge controller asks for higher prices than the PWM type but has wider applications and in a larger solar project, the mppt controller is the only option to work with.

MPPT vs. PWM: Pros and Cons

MPPT and PWM Charge Controllers perform the same tasks in the solar power system. we create a list that includes all the advantages and disadvantages of both technologies for your reference.

MPPT Charge Controller Pros

- Maximum Power Point Tracking Algorism Increase Power Conversion rate up to 99%
- 4 Stage Charging is healthier to battery
- Scalable for Large off-grid Power system
- Available for solar systems up to 100 Amps
- Available for solar input up to 200V
- Offer flexibility when system growth required
- Equipped with multiple protection
- Rich modes for load configuration
- Some can Charge Lithium (Lifepo4) Battery

MPPT Charge Controller Cons

- High Prices (usually cost twice of a PWM Charge Controller)
- Larger Size than a PWM regulator

PWM Solar Controller Pros:

- PWM Regulator has mature and proved techniques
- PWM Regulator is Simple Structure and cost-effective
- Easy deployed
- Less budget for a small project

PWM Solar Charge Controller Cons:

- Low conversion rate
- Input voltage must match battery bank voltage
- Less scalability for system growth
- Less Load Mode
- Less Protection

MPPT vs PWM Charge Controllers: MPPT Advantages

To charge the battery, the output voltage of the solar panel must be higher than the input voltage of the battery. If the solar panel output voltage is lower than the input voltage, then the charge current will be close to 0 (zero).

Moreover, the output of the solar panel is not a fixed value, the curve changes a lot during the working time, many factors like sunlight intensity, ambient temperature, even the dampness can affect the solar conversion.

When using with a PWM Controller, as the PWM system lacks flexibility, the setting parameters cannot be changed, that means during the charging process, the output voltage and current remain constant. even the solar input changed, nothing changed to the controller output.

as environmental factors like sunlight level, the surrounding temperature, and humidity are constantly changing. the energy generated from the solar panels may have wasted even the sun is good.

A PWM controller works like a switch and connects the solar array directly to the battery when charging. This requires the solar array operates in a voltage range that is typically below Vmp.

in a 12V solar system, the battery voltage range is typically 11–15V, but the Vmp voltage of the solar array is typically around 16 or 17V.

Since PWM Controllers don't always operate at PV arrays Vmp, this will cause energy loss, The greater the difference between the battery voltage and the Vmp of the panel array, the more energy wasted.

The invention of the MPPT controller solves this problem, by the favour of MPPT algorism, The controller automatically tracks the high power points of the PV to ensure maximum energy is obtained from the solar array.

in a solar power system, the cost of solar panels and batteries accounts for 80%-90% of the total budget. and the controller only takes the rest 5%-10%. but If you choose the right charge regulator type, a 5% budget solar controller can optimize the system to its best performance. take an mppt type is highly recommend when you looking for RV solar solutions. check the guide here.

Here is an example for you to better understand the mppt advantages, let's say you have a 1000W panel system, if you replace the PWM controller with an MPPT type, you only need to install 700W solar panels to get the same power.

the solar panels price is about $2/W on the market, then the whole cost of solar panels can be reduced to $650. In such cases, in larger systems, the savings can be greater, including the cost of purchasing the panels, cables, and more.

Besides, there are many amorphous silicon solar panels in the market. One feature of this panel type is that the open-circuit voltage is high, and the current is small. this kind of panel works better with an MPPT Controller.

MPPT vs. PWM: Circuit Board Differences

Conclusion: Which Charge Controller Is Best for You?

There you have it-an MPPT vs PWM showdown on a feature-by-feature basis. the mppt type is the winner in my eyes, but the PWM type still has spaces.

Based on everything you have seen, here is my conclusion:

MPPT Charge Controller is best for professional owners who need one controller that accomplishes heavy tasks (Home Power Supply, RV Solar Power, Boat, Hybrid Solar Power, and off-grid power station). Because of its robust features, it can save you money by cutting other expenses.

PWM Charge Regulator is best for small off-grid power applications that don't need any other features and has not much budget. If you just want the basic and economical charge controller for a small lighting system that does not try to fit the needs of others, the PWM controller is the way to go.

Chapter Six

Voltage Stabilizer

Stability is always good, and it is safe for that stability and does not let them cause any problems due to unrestricted or unwanted fluctuations. If you live in a place where volatility is normal and still does not choose to invest in good stabilization, then I will tell you that you can lose your home in the most expensive goods house.

Also, there are many cases which prove that voltage fluctuations may be extremely dangerous during the intervening period, but we have seen minor explosions due to volatile voltage. The only way to stabilizer is to regulate the voltage stabilizers valve and keep them in the correct range, so that we can choose the best quality.

The fundamental principle of the operation of voltage stabilizer

Voltage rules are required for two different purposes; The process of raising the voltage under voltage position under voltage position is called float operation but reducing voltage from overvoltage status is called bank operations.

These two main operations are required in each voltage stabilizer.

As discussed above, the components of voltage stabilizer include transformers, relays, and electronic circuits. Stabilizer reduces voltage drops in incoming voltage, enabling electromagnetic relay so that it adds more voltage from transformer so that the voltage will be compensated. When the incoming voltage is higher than normal, the stabilizer activates another electromagnetic relay, such as reducing the voltage to maintain the normal value of the voltage.

How to choose the right stabilizer for your home?

Voltage stabilizer depends on device ratings for dimension stability, so when buying voltage stabilizer, the first and foremost thing is the power of all the equipment (or special equipment) that power supply through a stable device, usually quoted in the VA or KVA and must be considered. Power ratings of devices; If the power rating is not available, then calculate the power ratings voltage and the output of the current product. It is always advisable to consider the true RMS voltage of the load.

Another important factor is for the future. Therefore, a future extension needs to be set to determine total power ratings, for a specific period, the actual energy requirement is 20% more than necessary. Home needs, 200VA, 300VA, 500VA, 1KVA, 2KVA, 3KVA, 4KVA, 5KVA,

8KVA and 10KVA rated voltage stabilizers are appropriate. For industrial and commercial purposes, high power rated service stabilizers is needed

How does the voltage stabilizer work?

Now let us go into deepness and discuss how these stabilizers work the stabilizers are automatic and safe voltage regulators. First, they stabilize the utility voltage in your home and then go to any electronic device. It regulates the voltage coming from internal supply and then distributes it. According to the process, the voltage was high and low. How it is done using electronic circuitry so that the stabilizer changes the inbuilt transformer with the electromagnetic relay and coordinates the voltage right from it. The whole process builds a safe range, stabilizers and protects the utility. Its function is to test and stabilize excessive energy fluctuations.

How to choose the right stabilizer for air conditioners?

The answer is that you must do some research before buying a voltage stabilizer for the air conditioner. This means that checking the best stabilizers available in the market, researching the benefits and discretion of each stabilizer, discussing with users buying this device and asking the businessmen is also important. When you buy for stabilizers, the three main areas you need to remember always including the energy usage of the equipment, the nature of the equipment, and the voltage fluctuations in your city.

Also, knowing your place of air conditioner's placebo is an important thing in which stability is needed, especially the ratings that are mentioned in KVA, Amps, or KW format. And finally, you need the common versions and voltage needed.

Benefits of Voltage Stabilizer

1. Restricting equipment breaks:

Device refutation with voltage flux is normal. The rebuttal of this device can cost you twice as much. Servo Voltage Stabilizer is a one-time investment but provides better returns over a long period of time.

2. Protects the industry from damage:

The voltage stabilizer must have a number of electrical devices. This can also have a small voltage fluctuation affecting human life. Voltage Stabilizer controls the voltage supply on each device and prevents any such situation.

3. Home Saviour

Not only the industry, it is necessary to apply for the protection of family members. This protects your family from any type of damage

4. Less use of power

The voltage stabilizer controls the voltage supply, thus ensuring less power consumption. This means that at least, the device, which is the normal function of any device, gives the power of the power instead of facing the actual cause of damage to the device.

5. Decrease in Power Bill

As mentioned above, the voltage stabilizer gadget is used to control the power of devices. In fact, it is a method of applying for a home.

6. Low maintenance costs

The voltage stabilizer has a powerful motor which requires less maintenance. It is not only savvy,

7. Easy machine movements

The system stabilizer with a voltage stabilizer is a valuable time of industry that is easy to move whenever you need it.

8. High Durable

The voltage stabilizer can manage even the worst conditions and works the same way. Such durability helps to reduce voltage fluctuations and risks related to it.

9. Easy customization:

There is easy customization at the strength and different voltage rates. Moreover, its motor can easily change if it is damaged or stopped working.

10. Extremely careful

Applying a universal voltage stabilizer, spending too much in home or industry, because they cost a lot of costs, buying new equipment, repairing repairs, spending repairs, good electricity payments etc. Payment etc.

So, do you want to give all the electrical equipment such as AC, refrigerator, cooler or industry without the support of the servo voltage stabilizer? Do not worry because I am not sure if it is expensive. The ubiquitous voltage stabilizer at any voltage rate is completely in the buyer's budget. Early

Tips for Buying Best Stabilizer for Air Conditioners

Look at the power ratings, voltage, and the current look of your AC. Generally, in India, the standard voltage is 230 VAC and 50Hz. Remember this stability

The high flow of the device

Analyse the energy fluctuations in your area and then go to the voltage stabilizer that will carry all the unwanted streams.

Good features such as mounting, time delay system, indicator, extra load protection, and digital features will be good

We need to appreciate that UPS requires a stabilizer than would ensure that the input current is stepped up to the required voltage, this would improve its operation. In Nigeria we usually experience fluctuations.